I0555107

This book is dedicated to YHWH and my family: Joe 6, Joe 7, Kat and Janelle and spouses, and progeny.

With special thanks to Sister Melissa and Sister Sarah who got the book process rolling.

ISBN 979-8-218-15330-4

Copyright S.K. Fox, 2023

All rights reserved.
No part of this publication may be reproduced, stored in any form whatsoever, or transmitted in any form or by any means without prior written consent from S.K. Fox.

Written by S.K. Fox
Pictures Contributed by S.K. Fox

Published by
Debarim Publishing, LLC
807 W Broadway
Spiro, OK 74959
www.debarimpublishing.com

Table of Contents

CHAPTER ONE

Before you Start Homesteading

Get out of Debt

This is SO important because being in debt will make the whole homesteading experience much harder. Unless you already have the land, equipment, livestock, seeds, fencing, and so on, homesteading has many costs. Having some cash available for these items makes the entire process smoother. To keep the costs of these essential purchases low, you can buy much of what you need used at yard sales or auctions. A wise move is to research your intended purchase before you buy. That way, you can judge what condition that item is in so you don't waste your money.

Sell everything you don't need

Start with that big-screen TV. When in neo-pioneering mode, you won't have time, desire, or even power to watch TV. Your new reality show will be your life. I urge you to keep a blog, vlog, or start a YouTube channel or diary about your experience. Even though many people are getting into the green, off-grid, independent homesteading life, each experience is unique. We all have the wisdom to share as we become part of the bigger homesteading community.

I shouldn't have to explain it, but fancy clothes don't survive the homesteading life very well. Your sheep don't care if you wear a Ralph Lauren pantsuit.

Lots of furniture. You will just need the basics: a bed, table, chest of drawers, and shelves. Those sturdy wooden shelves are more valuable than a leather couch in this lifestyle. You can have both but look at your circumstances before you buy or bring that double La-Z-Boy. It wouldn't fit in the 8' x 17' foot trailer we lived in for 2 1/2 years. Not everyone starts as small or rustic as we did, but many do, hence neo-pioneering. Imagine getting your belongings into a Conestoga wagon. You can always bring the leather couch onto the homestead later when you have built your home and/ or storage buildings.

Some Things You Will Need

Tools are your friend. They make your work easier, so you will learn to appreciate a quality tool. Get power tools if you have the means to supply electricity. Some people have a power pole installed and hook-ups available. Some people use a generator for electricity. If you want the authentic pioneering experience get old-school tools too, like a peavey, wedges, and a froe. Learn how to clean and repair your tools. If you are splitting a log in the middle of your woods and the ax handle breaks, you don't want to wait two weeks while the local handyman or hardware store service person fixes it.

Some tools are too big or expensive, and you may have to rent them. Here are a few tools you might need, especially if your land is undeveloped:

- chainsaw
- wood splitter
- sawmill
- rototiller
- skid steer
- bulldozer
- backhoe
- excavator-$1500 a week to rent
- tractor
- plus all the hand tools: shovels, axes, hoes, rakes, saws, etc.

Canning Pots, Water Baths, Pressure Cooker, and Jars

These are other homesteading tools of the cooking/preserving kind. Having them before harvest season makes processing what grows in the garden or orchard quick and timely. Canning jars have three parts-the jars, a small flat lid with a rubber lined edge and a metal band that holds the lid in place. New jars have all the parts included. Used jars may not have the lids or bands, so buy new lids. Used lids will not seal, and the rubber lining will be grey, not red if the lid has been canned before. Used bands can be reused as long as the band is not rusty or dented. I suggest getting different size jars, from gallon pickle jars to 1/2 pint jelly jars. If you are homesteading, gardening, and trying to be self-sufficient, you will grow a lot of food. What you grow, you will want to save. Canning is a great way to process much of your food-meat, fruit, dairy, and vegetables. Buy the *Ball Blue Book*. It is an

2

excellent beginner canning guide.

Animals and Gardening

If you know nothing about animals or gardening, research both. Both need basic information to successfully get your garden, and your herd/flock started. You can save yourself time and money by starting right. These two homestead staples are the basis for being self-sufficient with your food which is a homesteading goal for many people.

Valuable Resources

These are all-encompassing resource books that can tell you how to harness a horse team, grow grapes and make wine, help a nanny goat who's kidding, make cheese, companion planting in your garden, and so much more. *Back to Basics by Reader's Digest, Encyclopedia of Country Living by Carla Emery*, *LL Bean Game Cookbook*, the *Foxfire* series, and the *Storey* series contains a wealth of information for homestead living. The BBC shows *Elizabethan*, *Edwardian*, *Victorian,* and *War Time Farms* are full of old tools, animals, cash crop ideas, and methods for making a farm work. I highly recommend you watch some episodes when doing your homesteading research.

CHAPTER TWO

The Land

Land Size and Money Spent

Cheap land is around $1000 an acre. Any less than that, you need to take a hard look at what is being sold and why it's so cheap. Banks often won't give a loan on raw land, so be prepared to pay for your dream piece in cash or work an owner financing deal. If you can't pay the total price but have some cash, make a big down payment to reduce the total you owe while making payments on the rest. Agree to the lowest loan rate you can for the shortest time possible. Most mortgages are for 30 years. The longer the time, the lower the payment, but you are paying huge amounts of interest over that time. Depending on your interest rate, you could pay more than double the original price. We paid off our land in less than five years by making a 40% down payment and double and triple payments until we paid off the balance due. If you cannot do that, at least send a small amount applied to the principal, the total due, with each payment. Paying extra towards the principal carves down the total amount you pay.

Questions to Answer Before Moving

- What is the climate of the area you want to move to?
- What is the yearly rainfall?
- What are the weather problems in that area?
- What grows well there?
- What kind of soil will you be growing your crops in?
- Does the property you are interested in have radioactive or heavy metal waste?
- If there are poor soil conditions, are you open to ranching?
- Will you be ranching for yourself?
- Will you be ranching to sell your livestock?
- Do you want to raise specialty breeds like heritage cattle, elk, and angora goats?
- Do you want to raise meat or milk animals?

Check the Neighbors

Try to meet them. A bad neighbor can be a huge headache, even legal trouble. A good neighbor can be a huge blessing: loaning you equipment, giving you produce, and sharing advice about the area.

Research Water

This is essential knowledge for where you plan to settle. Water availability made or broke our choices for places to rent and live throughout our adulthood.

Some questions to ask or places to go:

- Does this county have water problems?
- Have there been any articles on this in the local paper? You can ask the city or county courthouse about meeting minutes to discuss water issues.
  - Do people have wells?
  - Do people have cisterns?
  - Are springs abundant?
    - A clue would be location names with the word spring, like Blue Springs Creek or Cold Springs Lake. Also, certain maps have spring locations.
  - Are there many rivers, lakes, and ponds in the area?

County Courthouse, Land Appraiser, Property Tax Office

This is a perfect place to find more information about how much the county values the property. What taxes were paid on it in the last year? What was the price it sold at the last time someone bought it? What other facts will help your purchase decision? Sometimes the real estate listing and realtor have this information too. It's the sign of a professional realtor. However, many realtors don't have that information, and you have to do the research yourself if you want up-to-date facts about the land.

Be Prepared for your New Lifestyle

You can do many things before you move. You can watch television shows, documentaries, YouTube videos, and read books on homesteading. You can take classes at local parks, farmer co-ops, granges, community colleges, or agricultural schools. You can visit a living history museum, observe, and ask questions. You can try your tools, dig a ditch, try Back2Eden gardening on part of your lawn, plant vegetables, volunteer to prune an orchard, or build any other skill you will need on your homestead.

Learning new skill sets will save you time and money. You will have less of a learning curve and make fewer mistakes.

Before we move on, do not add any other bills to your budget. There is no need to pay for a Wi-Fi hook-up at your house if you are moving in two months.

Now, let's talk about your land.

CHAPTER THREE

Your Little Slice of Heaven

Spend time and effort deciding what you want as your homestead. Look at pictures, check out rural property sources like United Country Realty and Zillow websites, or drive around rural areas to see what homesteads you like.

Will your homestead have trees?

Trees add value to land for several reasons. Trees are a resource for heating, cooking, building, and a possible food source if there are nut or fruit trees or income if you want to sell lumber. Trees help with erosion, promote wildlife, and provide shade.

If you have a property with trees, you should check with the local fire department or county records to see if fires are a concern in your area.

Large trees around your homesite could pose a problem. This is why you should have already looked into the soil conditions of the property before purchase. Big trees may be uprooted in storms if the area has shallow soil. If a tree falls on your fencing, road, or buildings, do you have the tools and know-how to remove it? Do your neighbors, and are they willing to help? What will you do with the timber once it has fallen? Mill the log for lumber, chainsaw it into splittable sections, carve a bench out of it, or bark it for your cabin.

Will your homestead be pastured?

Pasture is a valuable resource because you can support large livestock on it. Pasture also requires maintenance like mowing, tree clearing, and fencing. Some considerations for pasture are:

Different sizes of pasture require different equipment. A few acres of pasture may only need a mower, but 10+ acres of pasture may require a tractor. Do you have the appropriate equipment with attachments for occasional mowing, bush hogging, etc.?

What kind of grass is the pasture? There are many kinds of grasses, and not all are good for livestock. Foxtail, Johnson grass, Pirelli, and others are noxious weeds that are sometimes illegal and harmful and steal the nutrients that other more beneficial grasses like Bermuda, Timothy, Brome, and Kentucky Bluegrass need. Talk to the local extension office about the pasture and grass in your area.

Basic infrastructure like fencing and gates is a necessity when raising livestock. When you are looking at a potential property, look for fencing. Is the pasture fenced? Is the fence barbed wire, field fence, metal pipe, or pallets? Different livestock requires different fencing so consider that when looking at the pasture.

Gates in smart locations in the fields can make your life much easier. Having as many gates as you need also is a benefit. Having gates well placed in relation to the road, obstacles, stumps, washouts, etc., is something to think about. You want your gates to be in good shape, strong, and wide enough to drive a truck hauling a trailer through. Gates should be able to be locked securely, contain your livestock, and remain upright. Your fence will be attached to the gate, and a strong gate will keep your fence tight. A bad gate will sag, be difficult to open, and will let your animals out.

Accessibility to water for your livestock is crucial. Does the pasture have a pond? creek or spring? Keeping animals is much easier if you do not have to haul water to them each day. Having more than one water source is even better. Having cattle drinking continuously from one pond can make the pond septic, and if you put new livestock like sheep on said pond, it can kill the sheep.

Ponds can become places where harmful germs grow because the water is standing, not running. They can be treated with beneficial plants like cattails, hornwort, pennyroyal, and lily pads which oxygenate and clean the water. Vinegar added to livestock water can kill bacteria. I added vinegar to my duck's wading pond when they showed signs of a common eye infection.

Your county extension office will have information about area water issues and the county water board.

Ditches are important for rain runoff next to your road. Look at the lay of the land. Does it slope downhill? Are there ravines there already? How much area will be draining? Do you want to divert the water to a pond or lake? Do you want to slow down the flow of rainfall with swales? You can build your pond by renting a backhoe or tractor with backhoe or track hoe. You can dig your ditches with a backhoe and line the ditches with large rocks to prevent erosion. Planting certain vines or grass can also help with erosion.

Will your land be rocky?

Is your land in a mountainous or rocky area? Rocks are a resource. With rocks, you can make fences, buildings, raised beds, foundations, and more. However, rocks will be an issue if you have shallow or poor soil. You can have a garden on rocky soil with lots of nutrients, but you must compost or bring in the dirt.

Will your land already be developed?

Look at what is already on the property with the same thoughts in mind as you would the raw land.

What kind of roads are on the property, and how many? Are there grass trails, gravel, mud holes, chip and seal, or paved? Do they reach all over the land or just by the house and barn? Does anyone else have an easement or right of way?

Easements are common in rural areas. An easement is a legal right of someone to use part of your property, usually an existing road, to access their property, especially when their land has no other way into it. Easements are granted by going to court. You cannot block the easement. You can put a fence if you put a gate. You can lock the gate if you give your neighbor a key.

The same rules apply if a utility like a gas company, the water board, or the electric co-op has an easement. You own the easement; they have the legal right to access their power poles, pipelines, or meters on the easement.

Is there electricity at the property? Does the property already have a meter in place? When was the last time the electric bill was paid? How much is the monthly usage average? What was the most expensive month? Is the wiring new or done back in the 1940s? Does your county have permits or inspections for housing and electricity? Poorly done wiring is a fire hazard, so you want to ensure the electricity is safe and orderly. Chewed wires and overtaxed plugs are bad signs.

Do the buildings on the land have indoor plumbing? Is there a lagoon, septic, or county sewage line? Is there an outhouse? A composting toilet is an option in a house with no indoor plumbing as long as it is legal in your state.

Does the land have a barn? How big is the barn? Is it made of wood, stone, or metal? Are there stalls, a hayloft, or a tack closet? Does the barn have running water or electricity? A barn can be a big help for keeping animals; a place for kidding or calving, milking, and storing feed and tack.

If the property has a barn, keep in mind the new metal barns are not even close to fire resistant. The roof will start collapsing 10 minutes after the fire starts, so remember that when evacuating your livestock in case of a fire. An old wooden barn with heavy log beams will burn for hours before starting to collapse. An old barn can be dangerous if it's leaning or has signs of termite or wood rot damage. Many people like old barn wood for decorating their homes, so think of selling the wood if the barn is no longer viable.

If there is no barn, some temporary alternatives are a calf shed, which is a timber-framed, metal-sided, U-shaped building that you order from a shed company. They come in many sizes and average $1000. For more money, consider buying a tiny house or Tuff Shed, which will be more than a calving shed but come in many configurations and sizes. Tuff Sheds can have windows, porches, and lofts and can stand in as a house, tool shed, chicken coop, or another outbuilding if your land has none. A much cheaper temporary shelter can be built by bending a cattle fence panel and covering it with a tarp. Add chicken wire or wooden ends, and you can have a chicken coop, or goat milk shed.

If someone lives in the house and sells it to you, they can show you where the garden is. If you are buying an old farm, the garden is often visible in the yard as a patch with less grass, possibly herbs and perennials still growing. Old garden plots usually have decent soil since people needed to grow their food before supermarkets. Putting your garden there is a good bet.

CHAPTER FOUR

The Cheapest Homesteading Options

Buy a piece of marginal land. Marginal land can have several different qualities: clay or sandy soil, steep or swampy, no or bad roads, no utilities, full of junk, and amenities far away. A property filled with junk can be a bonus if you can haul the junk to a junkyard and get paid by the pound for metal, wire, and aluminum. Once you buy the marginal land, you can purchase a used RV or trailer to put on it. You will work hard to get land like this up to snuff, but it can be done. It is easier than you think.

How Off-Grid Can Help

Solar power can provide electricity for your marginal land homestead and cooking, and so can batteries, especially if they are rechargeable. A solar system can be bought used, 50 cents a watt is a good price. Solar power must be inverted from 12 or 24 Volts to 110 to run something like a computer or a lamp. The key to charging your computer is the correct kind of inverter, a pure sine wave.

The trucking industry has changed appliance availability by making coffee pots, coolers, lamps, toasters, and small air conditioners that will run off 12 Volt power. So check trucking stores for appliances to run off your solar system.

Solar Oven

Baking is no problem as long as you have sunlight; winter or summer, you can bake in a solar oven just like a crock pot. I have made many types of food, including cobbler, stew, chicken casserole, cakes, bread, and roast beef in a solar oven. You can make your solar oven if you want to avoid buying one. They are a black box with a glass lid and aluminum panels to direct the sunlight into the box.

Water Catchment

You can catch your washing-up water off the roof of your shelter. This water will have to be boiled for at least 5 minutes at a full rolling boil and then filtered through a coffee filter or paper towel before it is drinkable. Depending on your area, you can dig your well. Remember that a well hole is inherently unstable and will need to be lined with culvert piping, stone, or brick as you dig down, so it doesn't collapse on you. Once you hit the water, a pulley system will be needed to pull the water up to ground level. Also, note that water is eight pounds per gallon. If you can afford a well dug by a driller, do that for your drinking water.

Always test the well water before drinking it. Talking to your local water authority or county cooperative is a way to find out the quality of drinking water in your area.

Places like Nevada, Arizona, New Mexico, and West Texas are naturally dry, and rainwater is scarce. Wells are expensive and must go deep. Move to a place with lots of water and no water use issues.

Other Water Sources

Sometimes, local fire departments let you use their water. When we first came to our land, we asked the local fire department, and they said we could fill up with water from their spigot on the side of the fire station.

Artesian wells are spontaneous springs that sometimes are available to the public. Our area has one on private land piped to a public road, flows year-round, and is free for anyone to use.

Growing your Food

On marginal land, you have options. Raised bed gardening is one, and buying good soil will guarantee a healthy garden. Back2Eden is another gardening method using cardboard and mulch. Walmart, Dollar General, and many liquor stores have plenty of cardboard they will give customers. Tree trimming businesses, and sometimes the county, will possibly bring out wood chips to dump for free or will allow you to shovel their wood chips for free. Container gardening is another option for vertical gardening. Free soil can sometimes be had when someone drains a pond or does landscaping or construction work.

* * *

Pallet Building

Look in your area for free pallets. We have found them at construction sites, gardening centers, hardware stores, etc. You can use pallets for your non-skilled construction like compost piles, fencing, goat shelter, duck houses, and skilled building like bookshelves, outdoor seating, headboards, teacup holders, etc. The sky's the limit, and there are many ideas online.

Mortgage and Bills

When you pay for your marginal land outright, in full, you have no mortgage. When you are off the grid, you have no utility bills. Your costs will be solar panels, batteries, wires, inverter, charge controller, candles, flashlights, propane, and camp stove to provide for your basic cooking, lighting, and electricity needs.

Humanure Composting

This is an option if septic is not available. You will need a separate composting area and a ready supply of organic materials like peat moss, sawdust, leaves, coconut husks, straw, etc., to help break down your waste. This compost cannot be anywhere near your water or garden. Humanure is good for flower beds and trees and takes a year to break down into dirt that can be safely moved. On bad soil, the humanure will help you start your forest or hedgerow.

Get Chickens

Chickens are your best first livestock choice for your marginal land. They are cheep (I meant cheap. Get the pun?), easy to care for, provide you with eggs and meat, will clear the land, eat bugs and seeds, with a rooster and the right breed will reproduce, give an alarm to predators, and you can use their composted manure in your garden or to improve your soil.

CHAPTER FIVE

Moving

Once you've found and purchased your land, it's time to move. Moving is a huge life stressor and is a complex operation with many things to consider and do. I have moved several times across the country and internationally. Here are some things I suggest.

Economics

Moving will cost money. Budget for it.

- How much for the moving company or U-haul truck if you move?
- How long will the trip take? How long am I renting the truck for if I move?
- Is the truck rental one-way, or will you return it to the start point? One-way rentals cost more for the truck.

Other Logistic Concerns

- How many square feet of goods do I have?
- Can the rental truck also tow a trailer?
- Can it tow an animal trailer?
- Do you need a tow package for your vehicle or the rental truck?
- Do you have the lights package for towing a trailer?
- Will you need a ride to the truck rental location?
- What are the final payments of utilities and rent?
- What are the office hours of the utility office?
- Are there final fees or shut-off costs?

Change of Address

You should do a change of address form with the post office so they can forward your mail. Notify your utility offices and your landlord too. You may get prorated bill refunds and your security deposits back.

Don't Burn Bridges

This phrase means you don't have to leave with no friends or relationships intact. There is no reason to have bad relationships with businesses, churches, schools, or banks you dealt with before your move. They may have to vouch for you to set up an account in your new location.

You may need a statement from your old utility company before the new place okays your account.

Leave your old dwelling in a clean, safe condition. It would be easier for you to sell if you were the owner. It will be easier for your old landlord to rent it, and they will be likelier to give a good report of you as a tenant and not deduct it from your deposit.

If You Are Moving Yourself

Stash away free newspapers and boxes. Buy packing tape and begin packing before your move date. Take pictures or videos and list everything you own of value. Keep receipts of big purchases. You can claim damages to your property to the moving company and your insurance company as long as you have some way of showing what the item was and what it is worth.

Pack everything that doesn't move. Leave yourself one set of clothes and dishes for the final day, and pack them last. If you have a tribe, MAG, or church, let people know in plenty of time of your move date if you need help packing or loading the truck. If you have a moving company coming, check their inventory frequently to ensure all items are listed and what condition they are in.

Put heavier boxes on the bottom of the truck and lighter items at the top. Pack tightly, so there is minimal movement of boxes or furniture.

Arrival

Things to arrange once you get to your land/house/trailer/tiny house are:

- Who is helping you unpack?
    - Do you know anyone at your destination, or will it be just you?
- Where is your stuff going?
    - Do you need to rent storage space because you are moving into an expedition tent and don't have enough room?

- Budget money to pay for a storage unit.
- Figure out who is helping you move into and out of the storage unit, if anyone.

Some items needed for unpacking are a dolley or refrigerator shoulder straps for heavy items. You will also need a box cutter or knife to get through the innumerable amounts of tape and cardboard used to pack up all your belongings. Use opened, flattened boxes as welcome mats to protect the floor from all the helpers coming and going.

Keep your boxes and paper. Boxes can go in a Back2Eden garden, and paper can be used for wrapping meat at your next butchering session.

You are at your New Place

If you are on the grid, set up new utility accounts: electric, water, and heating. You will need to arrange trash service, get wi-fi hooked up, and open an account at the local bank.

If you are off-grid, get your well dug or set up your rain catchment system, water pumps, and filters. Install your wood stove or another heating source for your house/shelter/tent/yurt. Install your solar setup or plan for the space where it will go.

Now you are Moved in

You have basic needs taken care of, here are other ideas for you to consider. If you are writing, blogging, or vlogging about your experience, keep a daily journal of all you are doing. You can leave it to your grandchildren, if nothing else, and may be able to supplement your income by having a YouTube channel. You could write magazine articles or a book about your experience of returning to the land. If you work, plan your projects for your free time.

CHAPTER SIX

Animals

I got so excited getting ready to talk about the homestead animals I left out a very important concern for the critters: shelter. This is a concern that should come first.

Shelter/Barn/Shed/Coop

Barns and sheds come in all shapes and sizes. I love driving down country roads looking at old barns with their worn wood, enormous doors, haylofts, and so on. I have seen Ramada hen coops and little tiny two-hen chicken tractors and goat play areas that rival children's playgrounds. To each his own as far as style is concerned, but for utility, farm buildings need to fulfill a purpose and shelters especially.

Security is super important for your homestead livestock because they are food for you, so can they be food for predators. The shelter must be strong, complete, and able to withstand the onslaught of winged, clawed, and toothed critters trying to fly, dig or push their way in. If you have to build your own, construct sturdy shelters with closing doors and windows that can be locked and have screens for warm weather.

Shelters keep your animals out of the elements to the extent that they stay healthy, so the shelter should be rain and snow-proof and at least three-sided, depending on the occupant.

Chicken coops need 18"x 12"x 12" nest boxes and 1 to 2 feet of roosting space that is off of the floor by at least 3 feet. Roosting dowels are generally 1" to 2" around. Chickens go to sleep when the sun goes down. Like us, they are drowsy when they roost. They will not fight off a predator. The chicken coop should have at least these dimensions and can be bigger. Imagine having 60 birds. That's 60 to 100 square feet of roosting space, or at least a 6' x 10' space just for roosting. That is not counting the nesting area.

A shelter for ducks doesn't need roosting spaces. Ducks rest on the ground, but they are awkward on the ground, so again, their shelter must

18

protect them while they are inside it at night and must be able to be locked. Ducks need soft, absorbent bedding underfoot, not concrete or gravel, as it will cause problems with their feet. Their natural habitat is water. Imagine how soft that is on foot. Ducks lay eggs, so an area with hay or wood shavings with walls like a chicken nesting box will work for the duck hens to lay in. Duck laying boxes are at ground level.

Geese and turkeys are around the same size when mature, and act like chickens and ducks, so make their shelter like a coop, only larger. Domestic turkeys will rest on the ground, but heritage breeds want to roost at night. Geese rest on the ground just like ducks.

Goats and sheep need at least a three-sided shelter. It doesn't have to be tall but must keep them out of the cold wind and rain. They are prone to respiratory problems if left in cold rain and wind. As for their security, domestic neighborhood dogs have killed more of my goats than any wild predator in my area. Their shelter needs to be able to keep dogs, coyotes, or mountain lions out. If you don't build a shelter like that, you may lose animals to predators. Imagine a shed like Dirksen or Tuff Shed, about 8' x 10' with windows and a door. That is a perfect goat or sheep shed for a small to medium-sized herd of 2 to 10 goats.

I have seen a shelter made from a metal carport frame filled with wooden walls and a door on the end for dairy goats. The milking and kidding areas were inside with stalls and electricity. I have seen a commercially sold calving shed finished on the fourth side with wire fencing and a door. I've used 275-gallon liquid containers emptied with a small door cut in the side, and my present shelter is plastic wall panels tied together in a teepee shape. A bent-over cattle fence sided with vinyl or tin panels is sturdy and can be finished on the ends with chicken wire for ventilation and a door with wood framing. The possibilities are endless, and the decision for your livestock shelters is totally up to you.

As for horses, cattle, and llamas, they have thick enough skins and hair to withstand a lot of weather and are too large for predators up to bobcat and coyote size to mess with. If you are in an area with wolves, mountain lions, and bears, you will need a strong, wooden, or metal barn that can be locked. A horse stall has to be 12'x12' for the horse to stand up after laying down and about 10' tall. So for large animals, your barn will be a significant cost to buy or build and will take up a lot of space.

Shelters should have room for equipment like a milking area for goats and storing food. I store feed in metal trash cans. Mice and rats can chew through most plastic. Possums and raccoons can push unsecured lids off. The metal trash cans keep mice from chewing, and if I need to secure the lid, I use bungee cords. I have a separate tack shed for horse saddles, bridles, blankets, etc.

An old chest freezer works for storing feed, but you are liable if a small child climbs inside and cannot get out, so keep that in mind as far as access is concerned.

Put some thought into what shelter your animals will require, and get that shelter in place before you buy your livestock, so you are ready to give them a safe home.

Most homesteaders have animals as supplementary food sources, horsepower, security, or business. Let's discuss some of the typical farmstead animals and their uses.

Pasture/Fence

I know I touched on pasture and fencing earlier in this treatise. That was more broadly as you planned what your land would be like. This is a short list of specific fencing for the types of homestead animals you may keep.

Barbed wire and T posts are seen and used everywhere in the United States, usually for large animals or to keep large predators out. Strong, enforced corners are standard, with poles every eight to ten feet.

The field fence has welded wire squares and comes in different sizes. Field fence is usually for smaller livestock like sheep, geese, ducks, and turkeys.

Hog panels or cattle panels are made of much heavier gauge wire and come in 16-foot by 4-foot lengths. I use them for my goats also but have seen where paddock and loading areas are made from these panels for cattle and hogs.

Chicken wire is a smaller gauge. Instead of welded wire squares, the holes are smaller and more oval-shaped so birds cannot slip through.

Chicks can sneak out of this fencing, so an even smaller size like screening is needed to contain them. This fencing is for chickens, ducks, and guinea run areas or cages. This fencing can be paired with a plastic bird net over the top of the chicken run supported by a ridgepole or other horizontal framework to make the run area truly secure. 4 x 4 wooden poles are joined by 2 x 4 lumber to make a complete framework with chicken wire attached to the frame to stop larger predators from breaking into the bird's run. This ensures no space for any predators to get inside. The chicken wire is buried a few inches under the dirt at ground level and banked outwards, so predator digging is also discouraged.

Electric wire or net fencing is becoming popular for small livestock, especially with newer varieties running off portable solar batteries and plug-in kinds. Two brands of reliable electric fencing are Premiere One and Gallagher. Your local farm co-op will also carry conventional electric charge boxes, insulators, wire, and grounding rods.

Wood posts with cross pieces, metal piping, and electric wire are some of the fencing options that can be used for horses.

Chickens

Chickens are some of the easiest to care for, enjoyable to watch, and useful livestock for the homesteader. Chickens live five to six years. It takes 21 days for chicks to hatch from fertilized eggs. If she does her job right, a broody hen will hatch chicks and take care of them until they are grown.

Chickens lay eggs every day or other day. Eggs are a reliable source of protein. Imagine if stores, fast food restaurants, and butcher shops are not working or have limited access, like during the COVID pandemic. Having your egg-laying hens is a kind of food security. Chickens eat bugs, including beetles, grasshoppers, and more. They also clear the ground. They scratch to find seeds, larvae, worms, and bugs. If you keep them fenced in an area, they eventually will clear that area of all plants and fertilize it simultaneously.

Chickens require a secure shelter, especially for sleeping at night. It should be closed off enough so no predator can get to the birds. The hens need laying boxes 12H"x18W"x 18L" three feet off the ground, not on the floor where snakes can slip inside. The boxes should have some type of soft organic material in them, like leaves, grass, or wood chips, to keep the eggs from cracking when laid.

Predators that want to eat your chickens (and other birds):

- hawks
- owls
- weasels
- possums
- raccoons
- bobcats
- coyotes
- neighborhood dogs

While they are sleeping, chickens go to the bathroom so the coop floor will accumulate manure. To absorb the manure, you should always put some kind of organic matter in the bottom of the coop, like grass clippings, straw, leaves, wood chips, or pulled weeds. Dried chicken droppings form dust that can cause respiratory problems. By putting organic material in the bottom, you make a mini compost pile, giving off some heat in winter and making good soil over time. Clean the compost out once you smell the ammonia-like odor, and replace it with more grass, chips, or weeds.

Chickens need protein and calcium in their diet. Most modern feeds have both. However, chickens can sustain themselves at least in the summer months purely by foraging if you have enough land. You can raise your chickens on your leftovers. They will eat almost anything, including hot dogs, pizza, cake, salad, fruit, nuts, french fries, dog food, and more. I have also witnessed or heard first-hand testimony of chickens eating frogs, small snakes, scorpions, and mice and picking bones clean, especially when they are free ranging.

I have always suspected chickens of being tiny velociraptors because of their propensity to clean a deer carcass to just bones. A recent trip to the Alligator Farm near Jacksonville, Florida, confirmed my suspicions. The world-renowned reptile farm had an exhibit in which they showed how with just a few minor skeletal tweaks a chicken is a reptile walking upright.

You need a rooster if you want your eggs to be fertilized and possibly hatched into chicks. One rooster can cover twenty hens. One of the benefits of having a rooster is security. Roosters doing their job will call an alarm when they see the danger. When a hen gives an alarm, he will run to her and, if needed, attack a predator. The rooster will crow every morning at 4/5 am. He will crow periodically during the day and in the coop if he hears a loud noise.

Guinea Fowl

An African bird that found its way to America, the guinea fowl is a common dinner item in Europe. I saw my first guineas in the Pacific Northwest while horseback riding. I rode my Arabian gelding down a country lane and surprised a flock that flew away from us and made a heck of a noise. I was intrigued by these birds, did some research, and decided they made great alarm birds. I've had them periodically for several years.

Guineas have found a place on modern homesteads for a few reasons. They eat bugs, including ticks and chiggers. They sound a loud alarm whenever they see anything new or different including predators, visitors, and equipment. They also taste like pheasant.

Care and feeding are similar to chickens. One caution, though. As guineas mature, their instincts are to roost in trees. You will want to train them to come to the coop every night by herding them in before darkness falls or bribing them with a little grain. Also, female guineas will want to nest out on your property and will most likely become prey to something. If you want guinea keets, keep the hens in a safe enclosure.

Guineas only lay eggs in the spring and summer. Their eggs are smaller than chicken eggs with very hard shells. Their eggs have better nutrition than chicken eggs and taste very similar.

Ducks and Geese

Waterfowl add another dimension to the farm since they will be quacking during the day. I find ducks and geese adorable, but not everyone does. They have unique needs, and if you treat them like chickens, you may have problems.

Ducks and geese both need water, lots of it, daily. A pond or stream is an ideal water source. A hose works too but keeps in mind chlorinated water may not be the best option. Waterfowl are not designed to process chlorine, which may cause health issues over time. If you have treated water, let the duck water sit for a day or two before exposing them to it. Many modern homesteaders do not have ponds, so a quick alternative is a kiddie pool. If you use a kiddie pool, keep in mind the bottom will get soiled by their bill cleaning and need to be dumped out periodically with fresh water. A plus side to the kiddie pool situation is that water is a liquid fertilizer and can be used in your orchard and your garden.

The fowl use water to wash out their bills and clean their feathers. Without it, they may develop throat and bill issues and be unable to clean and oil their feathers which would also cause problems. Fowl breed in the water, so leaving them on dry land may hinder their procreating. Ducks need to get off their feet, which are designed for swimming, not walking for extended periods. Neither fowl should be on rocky soil or concrete exclusively and have better foot health on grass and water. Fowl will also help "clean up" a pond with too much algae, fertilize the pond and seal the bottom over time.

It is best to use feed for ducklings instead of chicks, but as adults, most ducks can eat chicken feed. They can free-range and eat a surprising variety of plants, bugs, small animals, and water critters. Geese generally eat plants, insects, and small animals. Also, geese eat leaves, grass, and nuts in the winter but will benefit from grain supplementation like ducks.

Domestic ducks like Indian runners, Khaki Campbells, and Pekins will lay more like chickens, and their eggs are larger. Ducks eggs are excellent for baking as they have more protein and fat than chicken eggs. A cake with duck eggs will be fluffier and rise higher than chicken eggs. As the ducks' age, the eggs may taste stronger than in the first year of laying.

Geese lay much larger eggs. Domestic geese lay eggs only in the spring and for hatching, like many "wild" duck species.

Geese are often used as watchdogs. They, like guineas, will sound an alarm and challenge a predator or stranger with wing beating and calling.

Geese are used as food by roasting, and in Europe, they were the preferred bird for feast dinners in ye good old days. Both duck fat and goose fat are good for cooking. Geese livers are eaten as paste on crackers.

Ducks need protection from predators, especially when they do not have a large pond to escape. They can be trained to enter a coop at night and sleep on the ground. Geese also need shelter at night, although their main predators are coyotes and owls. Smaller predators will not attempt to grab a full-grown goose. Most domestic fowl no longer fly unless forced to.

Turkeys

I have only had turkeys for a short time, but friends swear by them. A grown turkey can weigh between 30-50 lbs. That's a lot of meat! While turkeys provide a lot of meat, there are some concerns you should be aware of. They are more difficult as young poults. I've heard they can drown themselves for at least the first three weeks if their water pan is too deep. That may have been an urban legend, though. You DO NOT ever give young poults cold water. It must be at least lukewarm.

Turkey poults require more protein than chicks. And must be kept warm, like chicks, until they are fully feathered and 6-8 weeks old. Once raised, they are hardy and tasty if you are so inclined.

Turkeys can be personable. Too personable. I have seen mature turkeys who follow their owners like dogs.

In spring and fall, the males are constantly thrumming, displaying, and gobbling, which can be unnerving to annoying. They also have bigger droppings than chickens; if they are parading around your porch, they can cause a mess.

There are several benefits to having turkeys. Turkeys eat mostly grass which makes them economical for people on a budget. They grow big when they are mature, which means lots of meat. The hens lay eggs at least in spring/summer, and heritage breeds may hatch poults.

As with most of your small livestock, you need to provide a safe place for them to sleep at night. Wild turkeys roost in trees. You will need a sturdy roost to support the weight of your birds in your turkey coop. Moms and poults nest on the ground.

Goats and Sheep

Goats and sheep are some of my favorite homestead livestock. Our goat journey began after a visit to a llama farmer. My husband and I researched llamas and were taken with the idea that llamas provide security, and wool can be used as pack animals, eaten, and also possibly milked. We walked up to the llama pen and instantly realized we had grossly underestimated the size of these critters. They were the size of horses with heads way above ours. Right then, we knew we weren't ready for llamas. We didn't have fencing, shelter, or tack for them; if we moved, we'd need a horse trailer. We thanked the farmer and left.

On the drive home, it occurred to me that we might find that goats could do a lot: milk, meat, pack animals, and weed eaters. Goats are way smaller and would require less infrastructure to keep. We made another appointment with a local Seventh-Day Adventist lady selling goat milk, and our goat journey began.

Milk Goats and Sheep

There are many goat breeds: Alpine, Saanen, Toggenburg, Nubian, LaMancha, Oberhasli, Kinder, and so on. Most dairy breeds can be big girls ranging from 60-120 lbs. Alpines have horns. Oberhasli, Toggenburg, Nubians and others do not. I like the horns because it gives the goat some defense against predators.

Dairy must be bred and have a kid to produce milk. Most breed in the fall. Goats/sheep gestate at five months. A few sheep varieties can be used for milk and produce more milk for longer than meat or wool sheep breeds. East Friesian, British Milk, Awassi, Assaf, and Chios are milking sheep varieties.

You can make yogurt, kefir, and cheese from goat/sheep milk. Yogurt requires a constant temperature, and usually a starter of raw, unflavored yogurt. Kefir requires grains, and since I don't make kefir, that's about all I know about it. Except it has excellent nutritional properties and is good for you. The grains can be gotten at a health food store or grown from other kefir. Cheese is a wide-open topic with lots of options for the home dairy person.

Milk separates into curds and whey. You make cheese from the curds and can make ricotta from the whey. Curds are the white solids that float on top of the whey liquid when the milk separates. Milk separates when rennet and the right temperature are used on the milk. Whey is cloudy and full of nutrients. The protein added to powdered supplements for weight lifters is made from whey.

Rennet is used to turn milk into curds and whey. Rennet is from the lining of the young animal's stomach. Rennet can also be bought at Caprine.com or check with your local co-op. Vegetable or organic rennets can be made from nettles, thistles, and other plants. I have also had goat milk turn to curds and whey sitting on the counter in the summertime.

Feta, Ricotta, Mozzarella, and other soft cheeses are made from plain yogurt, salt, and rennet. The process includes boiling and adding vinegar and other ways that do not require adding bacteria. I have successfully made feta and ricotta, adding flavors like dill, basil, and hot peppers in the brine to give the feta additional yumminess. I use a recipe from Fias Co Farm for making feta and have some in brine right now in my refrigerator. Soft cheeses are usually not aged as hard cheeses are, but a cheese press can be used in forming the curds.

You must make hard cheese like Cheddar, Colby, Gouda, Havarti, and others with specific bacteria that you can get online or from a natural food store or farm co-op. Making the hard cheese is similar to soft cheese, with the milk being treated with rennet, reaching a specific temperature, the milk forming curds and whey, and pressing the curds. However, in the process, there is adding bacteria powder. Hard cheese can be smoked and aged by hanging from cheesecloth or being covered in wax for some time.

Meat Goats and Sheep

Boer, Kiko, Spanish, Barbari, Savannah, and Pygmy are all meat goats. Meat goats generally put on more weight than dairy breeds. Many of the meat goats were developed in Africa, where grass can be scarce or of poor quality. Goats are nibblers like deer. Meat goats breed year-round, and as I experienced this past year, can have kids twice a year. Pygmy goats and older nannies of other varieties can have up to 3 babies in one kidding; more meat on the hoof.

Goat meat can be fatty if the goats are given grain and taste great as curry, barbecue, or stew. Goats eat many weeds, leaves, and grasses. Young goat meat, still milk-fed, is called Cabrito. Adult meat is Chevron.

There are many varieties of meat sheep-Katahdin, Dorper, American Blackbelly, Blackhead Persian, Dorset, and Hampshire. Sheep eat many types of grass, with the occasional weeds and not so many trees. Sheep graze more like cattle.

Sheep meat is called lamb for ages 6-10 weeks old. Spring lamb is 5 to 6 months old. Six months to a year is yearling mutton. Mutton is beyond a year and tastes stronger.

Our first self-butchering event was after Pastor Joe, still on active duty, shot a deer on a military base. We were too cheap to pay for commercial butchering and had no idea how to start the process on the deer he brought home.

A bit of advice: practice butchering. Buy a whole chicken and cut it into pieces. Buy a ¼ of a cow and butcher it.

We had the Foxfire series of books. The series are stories from elderly people in the Appalachians, telling about the "old days." There was nothing in the glossary about butchering venison, but there was pork butchery, so we followed the directions, and it worked fine. We cut up the deer following the instructions in Foxfire. We wrapped the meat in plastic wrap and packing paper and put the meat in the freezer. Foxfire books had saved the day for us.

After you finish butchering, you will have leftover skin, feet, head, organs, and intestines. If you live in a city, it may be a problem disposing of the offal. It will begin stinking in two days and have maggots in three. If you live in the country, you can feed the leftovers to your dogs, pigs (we don't keep pigs, but some people do), or coyotes by leaving them at the edge of the property or in your woods. You can also bury the offal, but some enterprising wild critter or neighbor dog may dig them up.

Wool Goats and Sheep

Goats with wool are Angora and mixed breed Pygora and Nigora goats. Many types of sheep have wool-Merino, Suffolk, Leicester, Rambouillet, Dorset, Romney, etc.

Shearing in America is usually done in the spring, before hot weather. Commercial shears are the easiest and most effective way to shear.

I have used scissors, sharp knives, a pocket knife, and old metal shears and all were very difficult to use. Once shorn, the wool is cleaned by picking out burs, twigs, and knots. Then it is washed in clean water, dried, and baled.

Once an animal is butchered, its hide can be tanned to make soft, lightweight leather which is very good for making gloves.

There are downsides to keeping goats or sheep, and you want to be aware of these issues before deciding on your livestock.

Goats can be very destructive in their nibbling. Large dairy goats like roses, fruit trees, and much of the garden as food can decimate the plants in your yard and garden in just a few hours. The same for sheep, but sheep are not as aggressive at ruining fences. Sheep and goats do better if their grazing is rotated; every few days is best. If they stay longer than two weeks in the same pasture, the chances of worm infestation increase.

Goats also can be very hard on conventional fencing. They will slip through 5-strand barbed wire. Field fencing wears down by standing on the cross pieces and pushing the metal poles down. The chicken wire must be securely attached to wood or metal fence posts and cross pieces with the clear understanding that the goats will stand on their hind legs and rest their forelegs at the top of the fence. We found that hog or cattle panels work well, at least four feet tall, sixteen feet long-lashed to trees, or sturdy fence poles. With largely determined goats, you might have to put a panel on top of another to make it 8 feet tall. The panels have very stiff wire, are strong, and are hard for the goats to push over or apart. Goats with horns can get their heads stuck in the squares until you show them how to turn their head to get their horns through. Only twice have I had to cut the fence square to free a goat. Panels work for sheep too, but so do other types of fencing.

Electric webbed fences are popular right now, and reports are that they work. Barbed wire is not good for sheep as they have easily torn skin, and their wool can get caught on the barbs and trap the sheep. In Wales, fences are made of stone and are high enough that the sheep do not jump over. A stone fence for goats would have no outcroppings and be very tall, as goats love to climb rocks.

Both goats and sheep can and often do get intestinal or liver worms like horses and cattle. Some breeds are more resistant, but knowing your animals and keeping track of them is the best way to keep them healthy.

Some signs that you may have a worm infestation in your herd are a sunken stomach area(right in front of rear hips), white gums and/or tongue, always hungry but not putting on any weight, weak and less energy, dull and rough coat, diarrhea, swollen face/jaw-caused by liver bleeding, this is caused by the liver fluke, which will also be in the pasture. Liver fluke requires a special wormer, available online, and you should rotate your goats to another pasture or treat your pasture with Permethrin Spray. Permethrin is powerful and bad for amphibians and fishlife, so read all the directions on the container carefully and follow them. Sheep have copper sensitivity and cannot have too much copper in their feed or medicine.

Billy Goats and Rams

Billy goats and ram sheep are necessary for breeding your do to get offspring and milk. Unless a hornless breed, billies and rams will grow large, curved horns much wider and heavier than the doe. Rams often grow horns that curl near the head, around the eye, and ear. Billy horns often grow behind their head and can be 2-3 feet long, depending on the breed. Billies and rams sharpen their horns by rubbing them roughly on trees, fence posts, and other vertical poles. This process can kill the tree, knock down the pole, and wreck a fence.

The cattle panels take these beatings with aplomb. The lashings or fence posts are more likely to break than the panel. Especially when a doe is in heat, the male will be hyper-aggressive, and if you have two males in the same pen, they will crash heads and horns together until they are bloody. You must take great care during this time for your safety if you have a hyper-aggressive male. I have been knocked down and pushed around by a billy.

There are ways to avoid this behavior. You can bribe the male into another pen or another area. If you have to put your hands on the male, be sure there is something like a fence, tree, or ATV between you and him, so if he gets pushy, he will ram the object, not you. Not all rams/billies get aggressive, and how you raise them will be a factor.

One alternative to having several billies around after lambing/ kidding season is to neuter all male babies except one. A neutered male caprine is called a wether. He is just like a steer as opposed to a bull. He will get to full size and weight, and if you plan on using your herd for meat, wethers are a smart way to raise meat animals.

A male in your herd is also security for your does. A ram/billy will charge any intruders.

The male will be the dominant animal and will always butt others out of the way for the best feed and treats, so that is one reason why the male is moved when the females are pregnant or have offspring. A wether is a good companion animal for a billy who has been isolated from the females.

The adult male sheep or goat will periodically smell very strong and engage in behavior like urinating on their legs or beard. These behaviors attract the does but can be offensive to people, so keep that in mind when building the goat pen in relation to living structures or other areas you may spend a lot of time.

Due to their size, goats and sheep can be transported and housed in much smaller arrangements than llamas, cows, or horses. Goats and sheep can fit into dog houses and be transported as many as 10 in the back of a pickup truck, more if you force them. The problem with transporting these critters is their ability to jump. Having some sort of cage overtop of the pickup will work, or putting them into a dog travel kennel. I tie mine by their collars to a rope strung up by the pickup's cab. I tie them so the collar cannot slide along the rope. Because they are herd animals, they take comfort in being close together, do not panic, and cannot leap out.

Llamas, Alpacas, Donkeys, and LGD

Llamas are large animals that can be used for pack animals, wool, or guardians for smaller animals like sheep. They can be ornery. They use spitting as an offensive and defensive reaction. They also use kicking as an offensive and defensive reaction and can break bones with a kick. They require tall fencing.

Alpacas

They are similar to llamas but smaller. They are specifically bred for fiber. They also spit.

* * *

Donkeys

Donkeys are also used as security for smaller herds. Donkeys are their own breed. A mule is a cross between a donkey and a horse and is sterile.

Donkeys are used worldwide as work animals. They pull carts, plows, mills, and are ridden. Some cultures milk donkeys or eat donkeys. Donkey skins are also used as parchment. Donkeys are similar to horses and mules in feed and care.

Donkeys have several characteristics that make them different from other guard animals. They can be ornery and have large teeth. They have long memories and will repay bad treatment. They can get by on standard to poor pasture.

LGD-Livestock Guardian Dogs

Anatolian Shepherds are excellent guard dogs, but not for beginner dog owners. They roam an area of 5 miles square unless fenced and can be very aggressive with any type of animal that comes into their territory. A friend with an Anatolian pair watched them kill a trespassing dog more than once. They are large enough to scare off any type of predator and are serious guardians.

They also may get aggressive with people, including those who have visited before, and one cannot predict who they will take exception.

The males are generally more aggressive than the females. However, I have known a female dog who killed a goat brought to her herd to be bred. Anatolians are not great with newly introduced animals.

Great Pyrenees

Great Pyrenees are excellent guard dogs. Like Anatolians, they roam an area of 5 square miles if unfenced. They are aggressive with predators who approach their flock or herd. They may also get aggressive with people but generally are not as aggressive as Anatolians, and many bloodlines are people-friendly. The males usually are more aggressive than females, but both are excellent guard animals for your homestead. Predators are scared off by their size and have enough fight to defeat most predators. They tend to be heavier and shaggier than Anatolians.

Feed your LGD separately from your herd or flock to avoid food aggression.

Cattle

I have never kept cattle but have lived near them many times and had friends with cattle. Here are my cattle impressions.

Cattle are large animals with a lifespan of 18-22 years. Being 2,000 lbs may not seem like a big deal, but logistics are completely different for critters this size. Having one step on your foot is NOT like having a goat step on your foot. You will not be able to wrestle a full-grown animal to the ground or anywhere else unless you train them when young. Halter and lead rope training the calf is easier than waiting til the calf is fully grown.

Strong fencing is required. Cows can jump and push metal fence poles over to get to the greener grass on the other side.

Lots of pasture is needed, one acre of grass per cow. Three elk can feed in the same 1 acre or 6-8 sheep. Over winter, hay will be needed to supplement their diet, as winter grass has very little nutrition.

Medicating an entire herd of cattle will require lots of medicine, whether it be an antibiotic, dewormer, or powder to keep flies away.

Cattle need a lot of water, whether by a pond, stream, river, or well. Damp conditions are not suitable for most cattle breeds.

A bull is required to breed cattle. Bulls are like billy goats and rams, only much larger and stronger. Special care is required when dealing with bulls, including pens to enclose them for shots, worming, and other interventions. Many people use artificial insemination instead of keeping a bull on their land. Or taking their cows to a bull and paying for the use of the bull. Due to their size, some breeds can have calving problems, which often requires a vet's assistance.

There are several breeds of milk cows; Jerseys, Guernseys, Holstein

Friesians, and Swiss Brown. On average, a single milk cow can produce 7 ½ gallons of milk daily, with two milkings; one in the morning and one in the evening. Cow milk yields the same products as goat milk: butter, yogurt, kefir, and cheese, and in larger quantities. Be prepared to milk for as long as an hour, with preparation included. A milk cow requires training to stand still for the milking. They can be tied to anything from a tree to a truck to a stanchion to a headstall. One trick to keep a cow from kicking was passed on in a YouTube video by Fox Burrow Homestead-tie a rope around the cow's ribs. Cow geometry requires a big breath, but with the rope in place, the cow thinks it cannot take the breath and so doesn't kick the milk person or the bucket.

Cattle meat breeds are many too. Black Angus, Charolais, Red Angus, and Texas Longhorns are the most popular breeds.

Butchering requires heavy-duty equipment to hang the carcass: rope, pulleys, metal rails, chains, sharp knives, bone saws, a tractor, etc. There will be lots of meat, hundreds of pounds of it. Thought must be given to when and how to store the meat.

Canning is an option, but you will need lots of jars. You could need at least a hundred jars and lids/bands.

Freezing is a popular option. Have lots of butcher plastic, paper, and tape, plus markers to identify the cuts of meat in the packages. Make sure you have room in your freezer!

Another preservation technique is dehydrating or jerking the meat. First, you will need to cut the meat into very thin strips. Then, marinade with your favorite flavor. Dry either in a low oven (170 degrees) with the door slightly propped open to allow moisture to escape, in a dehydrator, or over a low smokey fire outside over a large rack. Jerking the whole cow would take several iterations over time. A deer yields 70-100 pounds of meat. A cow 10x that.

Salting in brine or alternating meat and salt is another old-school way to preserve meat. Corned beef is a type of salted meat.

Smoking is often done in conjunction with jerking. The meat is cooked in a smokehouse or metal smoker oven with wet wood providing the smoke over long periods. A smokehouse can be large enough to smoke a whole cow or more at a time.

Wild Animals

Population numbers and quality vary. On 20 or more acres, depending on where you are, there may be enough game to sustain your family. Different parts of the country offer a variety of wild game, from fishing, antelope, elk, moose, bear, black-tailed deer, mule deer, pheasant, ptarmigan, and so on. Hunting takes time, ammunition, a firearm or bow, fishing poles, and fees. Even with all this, it is still not guaranteed that the hunter or fisherman will return with anything.

The purpose of a homestead is to provide a reliable source of quality food through livestock, a garden, and an orchard. Spending the time and effort of hunting is spared. You could raise deer, elk, turkey, quail, and pheasant on your land to ensure a wild and domesticated population. You could also have a pond close to your house that you stock with fish.

CHAPTER SEVEN

Gardening

I've been gardening on and off for 49 years, starting when I was 11 years old. I returned to the United States after my father's death in Bangkok, Thailand. My mother moved us to live with my father's parents in Indiana. Although they lived in town, my grandparent's gardened, especially my grandfather.

His yard had over 50 rose bushes of all types and colors. He had a compost pile by his shed. He gardened for people in their yards and produced massive plants and fruits by adding 10-10-10 fertilizer and horse manure into his compost. When our family moved into our own house, our yard included a one-acre garden, and grandpa would drive out to rototill, fertilize, plant, weed, and harvest from that plot. I learned to love gardening while watching my grandpa tend our garden. I learned to can with my mom as she preserved what he grew.

My grandfather's love for gardening most likely started because he grew up poor and lived through the Depression. His first home as a married man was a log cabin.

Fast forward many years to the present-I can summarize my gardening advice in a few thoughts. Try different gardening styles and choose the one that resonates with you. I've tried rototilling, square foot, straw bale, Back to Eden, concentrated rows, and raised bed styles.

What works for me in my area (rocky) and climate (subtropical) is a combination of Back to Eden to produce soil where there were only rocks, Hugelkultur raised beds, and semi-covered beds. Because heat, critters, and bugs are the most significant issues in growing vegetables for me, the semi-covered beds keep the rabbits and deer out, plus break up our intense sunlight, and seem to deter pests. However, I diligently inspect my plants daily and squash any intruders or knock them off into a bucket of water to drown them. To fight weeds, I lay 4-6 mil plastic on the beds as early as the winter before and leave it on until warm weather comes and I begin planting. My goal is complete weed breakdown, and the plastic does the job. I find the soil beneath the plastic is dark and soft. Later in the summer, weeds from seed carried in the rain runoff or dropped by birds or crawling underground from their viney root systems will show up. Still, by then, the vegetables are well established, and I can pull up the baby weeds with less effort than the heavily matted mature weeds that result if no weed-killing action is taken.

During the pandemic year, I discovered soil improvement techniques beyond what I knew from square foot gardening- adding clay to sandy soil and vice versa, adding bone meal and blood meal. Adding peat moss, diatomaceous soil, and fertilizer(duh) worked wonders. I had been trying for ten years here to make exceptional soil, but none of my efforts worked as well. The best commercial fertilizer I found is 13-13-13 strength, and I just freehand broadcast it over and into the beds. The best natural fertilizer for me is rabbit manure. The results I've gotten with these style beds and techniques are the best since I've been off-grid, even with the late freezes and drought of this summer. I'm excited for next year's gardening season.

I also have a do-it-yourself greenhouse made from bent fence panels, concrete block base, and 4-mil plastic. I plant in it in early spring and have successfully grown peas, tomatoes, Brussels sprouts, herbs, cucumbers, and cabbage. For me, it's less of a bedding plant growing space and more a protect my brassicas space. I screened the bottom half and shade-clothed the top. This way, I could grow cabbage, broccoli, and brussel sprouts protected from too much sun, cabbage moths, squash bugs, and Harlequin beetles-our main pests. The greenhouse is approximately 20 degrees above the outside temperature, and in mid-December, cabbages, broccoli, brussels sprouts, rosemary, peas, strawberries, and oregano plants are all doing fine inside.

The Hugelkultur raised bed plus rabbit manure has produced a very large Swiss chard plant, Rutgers tomato plant, beets, and carrots. This style of high-raised beds makes weeding and pest control easy but requires construction and materials beyond what the semi-covered beds did. The other benefit of this style of raised bed is the wood at the bottom retains moisture, so watering is not a big issue-something to be considered when living off-grid.

The best advice for a garden. Do some research into your growing seasons and local soil, then get started tilling the soil and getting seeds into the ground.

Wild Edibles & Perennials

Knowing and cultivating wild edibles for your area and planting perennial varieties of vegetables and fruits are two other aspects of producing your food.

Curly Dock, Burdock, Plantain, Jerusalem Artichokes, Lamb's Quarter, and Smilax are wild edibles that grow nationwide and are tasty and nutritious. Buying a field guidebook for your new homestead area, checking with your local library, attending a master gardener class, or stopping by the county extension office are ways to discover the weeds you can safely harvest. Wild edibles add to your diet every year and help extend your pantry.

While I lived in Kansas, I learned about lamb's quarter from an elderly lady who was a friend of mine. She had red hair, a large smile and her name was Jewel. She struggled through the Depression with her family. She showed me the leafy green plant, and I never forgot it. I became obsessed with finding it on my property, walking in town, and even in vacant lots!

I have taught many people to identify lamb's quarter, even my grandchildren. I still remember how she told me the identifying feature is the top leaves will be a lighter, silvery color and slightly fuzzy. The leaves are spear-shaped and nubby on the edges. I still use Goosefoot, as it is also sometimes called, in my cooking. I dry the leaves in my solar dehydrator and add them to stir fry and stew. The fresh leaves are tasty in a salad or used on a sandwich as a substitute lettuce. Lamb's quarter is also excellent fodder for all homestead animals, from goats to chickens. I do not know if cattle like it, and I imagine sheep will eat it, but I am no sheep expert.

Luckily for us on our homestead, some plants fit both categories: both wild and perennial. These include passion fruit vines, blackberries, black raspberries, hackberries, black haw, and huckleberries. We also have hickory nuts, oak acorns, and walnuts. My suspicion is some of these bushes, vines, and trees were planted here hundreds of years ago, and we are just the latest people to benefit from this land. Black Haw, for sure, was not native around the Civil War. It was popular for making pies and puddings and has fallen out of favor due to the meat vs. nut amount of the berry.

Passion fruit, I discovered, has 126 varieties around the world, and many grow in the continental US. Our property has passion fruit that grows every year. I don't know if it was planted or wild, but I let it grow and harvest the fruit as it ripens. The first time I saw the vine with the lime-like fruit hanging off it, I thought it was some noxious weeds! I let my goats eat it! I suffered the folly of not knowing the edible plants growing on my land. My kindly neighbor figured it out by watching a video and seeing the crazy purple and white blossom. He told me that when the flower and leaves are dried, they make excellent tea. Now I eat the fruit from the vine as soon as it's ripe. The flavor is a combination of lychee sweet and lemon tangy. I look forward to seeing the vines appear every mid-summer and they spread each year.

As for perennial vegetables, my absolute favorite is the Walking Onion, also known as Egyptian Onion. The bulb grows a green stalk. The stalk forms small onions on top. When these baby onions get bigger and heavy, the stalk falls and plants the little onions. They grow bigger and shoot up stalks. Once planted, you never have to plant them again. Make sure you have room for them to spread.

Most other vegetables can be a kind of perennial if you buy and plant heirloom seeds. Heirloom seeds will produce the exact type of vegetable every generation. People save seeds from the harvest each year, preferably the best and earliest fruit, like carrot, tomato, and squash, to be the seeds they plant the next year. I am a lazy gardener, and because our temperatures are not usually extreme, I try allowing some veggies to ripen on the vine or plant and fall back into the soil. I think that since weeds and tree seeds do the same thing, the vegetable seeds will lay in the ground after the outside has rotted and will sprout again when summer comes and the soil warms. This method has worked. I have Russian Mustard that manages to survive year after year, sometimes just a few plants, because the garden area where they were started doesn't get the black plastic treatment. The mustard greens are over ten years old, and I use the leaves the same way I do all leafy greens-dehydrated in soups and fresh in stir fry or on sandwiches. The plants that come up from letting mature vegetables fall to the ground or you put in your compost are known as volunteers. I'd love to have a completely volunteer garden as I get older.

Good luck with your garden. May it feed you with healthy and tasty produce.

CHAPTER EIGHT

A Way to Organize Your Homestead

I took this idea from a historical model of the first millennia of Europe. The finer points are from a model of homesteading life on the American plains.

Year-Round

In the year 1000, daily life was planned around a yearly calendar and agricultural events like planting, harvesting, pruning, milking, breeding, butchering, and so on. The life of the land, with its farming and ranching, hasn't changed that much in the basics. You can plan your year along the same lines, following general ideas around the seasons and months.

The chart is a general list meant for Zone 7. Change it to suit your needs. We also do not do crops like hay, so adjust the schedule as you need for plowing, disking, tilling, combining, etc.

CALENDAR YEAR

Season	Month	Activities
Spring	March	Turn garden soil. Plant potatoes by March 17. Plant brassicas-broccoli, cauliflower, cabbage, and peas if not already done. Plant greens like collards, swiss chard, and turnips.
	April	Usually the month of Passover. Get the leaven out of your house and life. Turn the garden again. Pull any weed starts. Plant beets, spinach, onions, and lettuce. Watch your herds for calving, lambing, or kidding. Fertilize the orchard, spread diatomaceous earth or spinosads, or other bug control

		Fertilize the orchard, spread diatomaceous earth or spinosads, or other bug control measures in the orchard and garden. Fruit trees will blossom.
	May	Chickens will start laying as the days lengthen. Goats and sheep will kid if bred in December. Plant hot-weather veggies like tomatoes, cucumbers, peppers, summer squash, corn, green beans, and watermelon. Keep weeding the garden. Good time to start putting up wood for the next year, before the heat. Look for wild raspberries to ripen. Feast of Pentecost this month. Have a day of celebration to Yahweh.
Summer	June	Goats, sheep will kid, if bred in January. Full of bugs in the garden, use whatever means you have to fight them. Weed the garden. Trim grape leaves so sun and air can get to the bunches. Harvest peas. Use a dehydrator for any harvest, especially greens. Wild greens, lamb's quarter, curly dock, clover, wild lettuce, etc will also be up. Blackberries, strawberries, and peaches will be ripe. Plant winter squash and pumpkins. Breed cows.
	July	Pick and can any ripe produce from the garden. Blackberries will be ripe. Wrap grape bunches in knee-hi material or cotton muslin. Tie so bugs and animals cannot get at the fruit. Orchards will have apples and pears on trees. Make blackberry wine and jam. Fight bugs and weeds. Plant peas, spinach, and lettuce again.
Fall	August	Begin milking nannies that kidded in April or May. Offspring can be weaned.

		Dry herbs, plant winter crops like Brussels sprouts, turnips, and broccoli. Pick apples, and pears, if ripe. Make pickles from cukes. Can green beans, peppers, tomatoes, etc. Harvest garlic and onions. Wild passion fruit will be ripe.
	September	Apples, grapes, and pears will be ripe. Harvest grapes/make wine or jam or juice. Prepare winter quarters for animals. Buy hay or straw for winter feed. Rake leaves. Put the rest of the garden to bed. Allow chickens and goats to clean up old, dead plants and spread manure.
	October	Feast of Tabernacles, week-long harvest festival, and camping out to Yahweh. Butcher young animals raised for meat, smoke, salt, or dehydration. Butcher old chickens, goats, sheep, and cattle if not wanting to feed them over another winter. Gather pumpkins and winter squash. Plant turnips and garlic. Should have all the wood gathered and split by now.
Winter	November	Rake leaves. Use leaves as mulch in the garden and the coop.
	December	Put ashes from a wood fire, manure, and compost in the garden. Do indoor work like sewing, quilting, and writing. Hackberries will be ripe. Persimmons will be ripe after the first frost.
	January	Order seeds and chicks if necessary.
	February	Prune orchard and grapes. Spray for bugs in the orchard. Put sulfur powder down for chigger and tick problems. Plant peas. Lay mulch in the garden and orchard.

Weekly

Day	Activities
Sunday	Community Day, work day or projects on land
Monday	Fun day, trips off the land, museums, etc
Tuesday	Work on buildings
Wednesday	Work on garden
Thursday	Shopping, appointments, banking
Friday	Prep day, clean house, restock
Saturday	Sabbath

Daily

Time	Activities
5 AM - 8 AM	Wake up, eat breakfast, feed animals, vlog
10 AM - 1 PM	Work
1 PM - 2 PM	Lunch
2 PM - 4 PM	More work or home school or town trips
4 PM - 5 PM	Clean up, dinner prep
5 PM - 6 PM	Dinner
6 PM - 7 PM	Dinner clean up, final feed and water animals, lock up
7 PM - 10 PM	Evening activities, bathing, reading, etc

Daily- Working with Children

Time	Activities
6 AM - 7 AM	Go to job or school
8 AM - 11 AM	Cleaning, chores, home school
11 AM - 12 PM	Lunch
12 PM - 4 PM	More work in and around the house, appointments, shopping
4 PM - 5 PM	Dinner prep, homework tidy up, dad home from work

4 PM - 5 PM	Dinner prep, homework tidy up, dad home from work
5 PM - 6 PM	Dinner, family discussions of the day
6 PM - 10 PM	Finish homework, yard work, bath time, relax, family end of day prayer

CHAPTER NINE

Food Basics for the Homesteader

Food preparation- growing, cooking, and preserving- is central to life in general, and especially homestead life. The core of a desire to homestead is often linked with growing healthy food. A well-stocked kitchen makes those food tasks easier, more sensible, and safer. Here are some basics I recommend for any kitchen-homestead or urban.

Recommended Cookbooks:

- Miller, Bob. Amish Country Cookbook, and Vol II. Elkhart, IN: Bethel Publishing, 1986.
- Backwoods Home Cooking. Gold Beach, OR: Backwoods Home Magazine, 1989-2003.
- Showalter, Mary Emma. Mennonite Community Cookbook Favorite Family Recipes.Scottsdale, PA and Kitchener, Ontario: Herald Press, 1983.
- Any version of the Fanny Farmer Cookbook.

Tools for the Kitchen or Cooking

- Dutch Oven
- Solar Oven
- Wrought Iron Frying Pan/s
- Pie pans, bread pans
- Baking dish, casserole dish
- Metal bowls for mixing and cooking
- Colander (at least one, can have more)
- 2 large pots(at least one stainless steel)
- 3 small pots(at least one stainless steel)
- Plastic and wooden cutting boards
- Utensils
  - Wooden, metal, and plastic spoons/ slotted spoons
  - Knives-chef, butcher, filet, bread, skinning, paring, etc.
  - Metal salad tongs
  - Metal spatula
  - Metal ladle
  - Rolling pin

- Metal meat fork
- Sifter
- Grater

Staple food items for basic cooking:

- Flour
  - DIY alternatives: hard red winter wheat, rice, corn, barley, millet, quinoa, amaranth, curly dock, acorns
- Sugar/honey/stevia/dried fruit
- Baking powder-inexpensive, loses strength over time
- Baking soda-inexpensive, good for cleaning, cooking, bug bites
- Yeast-can also capture it from the air
- Salt-inexpensive, stores forever
- A solid fat like coconut oil, vegetable shortening
- Milk-store powdered/canned/shelf stable
- Think about how you are going to get milk after an emergency: goat, cow, sheep, buffalo?
- Butter-store canned or powdered
  - Think about how you will get butter after an emergency.
  - Goat, cow, sheep, buffalo

Canning Supplies:

- Large bowls
- Vegetable peeler or knife
- Cutting board
- Water bath canner-for high-acid foods like tomatoes
- Pressure canner for low-acid foods like green beans
- Jars/bands/lids
- Jar lifter
- Canning funnel
- Rubber gloves
- Large spoons/ladles *Ball Blue Book, Stocking Up, Reader's Digest Back to Basics* or any other book with canning instructions. Sure-Jell or any other canning pectin.
  - Pectin can be made from apples and other fruit.

Milking

- Milk stand with headstall
- Sweet feed or another treat for the nanny/cow/ewe
- Stainless steel bucket or pot
- Warm soap and water

- Dry towel

Steps for Milking Livestock

- Put the animal in a headstall.
  - small livestock on a milking stand.
- Secure headstall.
- Put sweet feed or treats in a milk stand tray.
- As the milker eats the treat, clean her udder with warm soap and water. Dry with a towel.
- With cows, grab a teat and stroke downward while also squeezing the teat.
- With goats/sheep, use your thumb and forefinger to close off the top of the plug-in and close the other fingers to force milk down and out of the teat.

*HINT: Do not tie your goat/sheep/cow like a trussed holiday goose when it's time for milking. You will teach your animal that being trussed is normal and milking is an ordeal. When you milk, simply milk until the udder is drained. Then, release the animal. If the animal fusses while in the headstall, continue to milk. This teaches the animal that milking is the purpose, and it's going to happen no matter the fussing. I have trained several goats to milk over twelve years, even those a human had never handled, and they all came to be docile in the milk stand.

CHAPTER TEN

Machines vs. Manpower

Nobody said to have a homestead you must be off-grid. Machines help you save time and effort. Here are a few suggestions for homestead helping machines:

- tractor
- log splitter
- chainsaw
- ATV
- 4 wheel drive full-size truck
- generator
- lawnmower

Manpower

If you want to be off-grid and go old school pioneering with hand or animal-driven tools, start shopping at thrift stores, antique stores or farm auctions. Know what tools you want and search for them. Old tools are usually sturdy and effective. Here are a few suggestions for useful hand tools:

- canning funnels & jar lifters
- apple peeler/corers
- hammers
- chains
- horse collar
- hames
- draw knife
- twist lid remover
- peavey
- two man saw
- breaker bar
- ax
- sledge hammer
- pole pounder
- pole remover

* * *

CHAPTER ELEVEN

Vehicle

What kind of land you buy will directly affect what kind of vehicle you drive. In rural areas, the highway departments may not be as fast on treating county roads for the weather. Floods may periodically shut down local roads. In cold weather areas, the occasional ice or snow will do the same. Here are some options to consider for your transportation.

Truck

A four-wheel-drive truck, whether crew cab or not, will get you through most driving situations. Bad county roads, rutted and graveled, will not cause nearly as many problems as driving a car on the same roads. Cars are designed for travel on level, paved roads. Trucks have heavier undercarriages and suspensions because of the expected usage. Trucks are better for carrying supplies like hay, grain, tools, kennels, fencing, etc. A four-wheel-drive truck will also be able to tow trailers or other vehicles.

Tractor

Someone suggested a tractor right after we moved to our property, but we were inexperienced, and the safety issues plus financial costs were more than we were ready for at the time. Once we used a borrowed tractor and saved the funds, we were more prepared for the equipment. I have a video of the tractor on my YouTube channel, Sister Kate Shofar Mountain. The difference for my husband around the homestead for heavy work was night and day. Now with attachments like a box blade for the roads, a bucket for hauling dirt, and chains for moving trees and rocks, we can get much more heavy work done around our land. That helps us finish the various projects we have efficiently.

All-Wheel Drive Four Cylinder Vehicle

A truck or car is acceptable for running to town, getting groceries and going to the bank. A four-cylinder car will use less gas than an eight-cylinder truck, and unless the roads are bad, save the wear and tear to your strong vehicle by having the smaller, more efficient scooter car for everyday driving. If you can get the scooter car in all-wheel drive, like a Rav 4, Jeep Liberty, and so on, even in bad weather, you can drive it on paved roads, and it will get up the steep, gravelly county roads too. You can use your truck to go pull people out of ditches when it snows.

ATV or Mule

To be clear, I mean the mule vehicle, a rugged golf cart that motorcycle dealerships now offer. The first versions were one or two-person only. Later versions are larger and can carry six people.

An ATV, an all-terrain vehicle, usually is built for one driver. Large versions can have room for one passenger. ATVs are built for off-road use, have plenty of clearance with large under-pressured tires, and have an engine size of 49 to 1,000 cubic centimeters. A 350 to 400 cc ATV can be rated to carry up to 500 pounds. That weight rating is the same for a small, four-cylinder Toyota truck. ATVs come in part-time or all-time four-wheel drive. On uneven ground, a four-wheel drive is a must.

An ATV makes homestead chores easier for me in moving square hay bales and bags of feed to our various animal pen locations. We drive on our land to farther areas when we take lots of equipment for projects like working on fences, clearing areas, and going to the range.

I recommend walking if you have lots of land because it's healthier. So, if time allows, walk instead of driving.

Old skool (forgive the spelling, I'm trying to be hip here)

You could have a team of oxen, draft horses, or mules. They require their harnesses which can be bought new from tack stores, the Amish, or thrift stores. Harnesses are a multi-part leather strapping system to attach the draft animal to you by the reins and do whatever you are pulling by straps or chains to a single or double tree. The parts of the harness are

- Bridle with blinders
- Driving reins, longer than normal reins
- Collar-the large padded collar you see on the Budweiser Clydesdales.
- Hames-the metal bar apparatus that fits over the collar and through which the reins go. The harness also attaches to the hames.

- Harness-should be complete and attached securely on the horse's chest, belly, and haunches. The harness should have straps to attach to the single or double tree.
- Single or doubletree-a wooden pole with hooks that attaches to the harness and has hooks for attaching a plow, wagon, etc. This is usually suspended between the horse and the wagon when pulling something.
- If you want to be organic or natural and use draft animal power, you are opening your choices to where on your land you go and how you use these animals to accomplish goals on your land.

Having a horse on our property allowed me to ride the perimeter much faster than walking while avoiding ticks and chiggers. Because our land is heavily treed, horseback was the easiest way to get anywhere on it.

Having draft animals means you can walk the animals to your work site, on your property, and begin tree felling, plowing, rock moving, or stump removal without installing a road first. Pulling out logs may be a different story. It would be difficult to pull a log up or down an incline with draft animals, but it would be doable on level ground.

Draft animals can be fed from your land if you have pasture and reproducible if you have a mare and a stallion or access to a stallion. Oxen are defined as certain breeds of cattle. Reproducing oxen will require a cow and bull of that breed. The good news about using oxen is they are way more edible than your tractor if things go sideways. Mules are sterile so you need a Jack donkey and a mare pony or horse to reproduce a mule.

A simple pick bought at a farm store will do for cleaning the hooves, and a pair of trimmers and a file for keeping the hooves from overgrowing. Find a local farrier if you don't want to do the trimming yourself. The Amish or Mennonites in your area are a good place to start looking for a farrier who can work on draft animals. You could also contact a local vet.

Science

Depending on what homestead you choose, science will be a factor. Studying and knowing something about science will make your homesteading experience easier.

A basic scientific principle for work is the weight of object times the amount of distance you will move it. When moving a bigger object, more force is needed. When moving a long distance, more force is needed. Knowing your strength will help you decide what work you can do and when you will benefit from having equipment. I know this sounds simple to some folks, but allow me some leeway for those coming from a non-rural background.

Let's use tree cutting as an example. Before mechanization, trees were cut with axes or saws. A person, let's assume a man, could chop down a tree with an ax. It might take him a half hour of solid work to fell a tree 20 inches in diameter. Now you have a basis for planning your work. If you are planning to fell ten trees today of that size, and you are an average man between 30 and 40 years old, you could estimate it will take you five hours to fell those ten trees. (That's some math right there.)

Now, let's modernize a bit. That same man with a chainsaw can cut down that same tree in fifteen minutes. We have to factor in that the chainsaw is heavier than the ax. Less work is used to wield a chainsaw since the chainsaw cuts directly and isn't swung back and forth. The ax is lighter but is swung back and forth many times. The chainsawing takes about half the time of the ax. In the same time of the ax, the chainsaw will do double the amount of work, cutting 20 trees in five hours.

Now, if you have land to clear, and your 1/2 acre has 40 trees on it, with the chainsaw, you will do 10 hours of work to cut those trees down or 20 hours with an ax. Of course, these figures are general and depend on many other factors, like the weather, how you feel, and whether the land is sloping, rocky, or wet. However, having general information helps in planning your work.

Weight and Gravity

Heavy things can be dangerous, even if the heavy thing is a rock and has no mind of its own to cause trouble. (Think of a bull.) When you lift a rock, you have added another consideration of how to move it because in lifting it, there will be force downward when it is dropped or let go. Picture that rock in the scoop of a tractor. The tip of the tractor loader affects the safety of people and objects around the rock. If the scoop isn't tipped up enough to keep the rock inside, the rock could fall out with even more force because you've lifted the scoop five feet in the air. I know that is a "duh" thought to some of you, but look at videos on Youtube of "OOPS" situations. You'll see all kinds of situations where people were working with heavy things and didn't consider these forces at work. By thinking about these forces before doing the work, you can avoid making mistakes like trees falling through houses or ATVs being loaded too fast on a truck and busting a rear window.

Another critical idea is that heavy objects will move downhill by themselves because gravity is acting on them if you give them the opportunity. So, firefighters are trained to put chock blocks under the tires of the fire trucks whenever the trucks are stopped at an accident scene. Why? Because fire trucks are very heavy, and will roll if the driver leaves the truck in neutral, if the ground is icy or if the road is covered in oil. So the trucks are chocked for safety to keep them from rolling out of control.

On a farmstead, the same law of gravity applies. If you are hand-digging a well, keep the sides square, so the top layers don't collapse in on you. If you cut down a tree on a hillside, make sure it won't roll downhill, especially if you are downhill from it. Ensure all vehicles are secured by putting the vehicle in drive on an uphill or reverse on a downhill with the front wheels pointed in a safe direction. If the vehicle is in neutral, setting the parking brake or chocking the tires will secure it. Never underestimate the force behind a heavy object. If it can roll, it will.

Lever and Fulcrum

Again, this is simplistic but very effective if used correctly. So, to lift a heavy rock, a lever and fulcrum can help you do work that you normally couldn't do by just lifting. A fulcrum is a wedge or rock that you place next to the heavy object, and the lever is a straight strong pole, like a breaker bar or stout pole. You place the pole against the rock and on top of the fulcrum. It allows you to apply more force than just lifting. People will do less damage to their backs if using these tools instead of trying to lift things that are too heavy for them.

Block and Tackle and Pulleys

Block and tackle and pulleys are wheel-shaped devices that a rope fits in and over that do the same basic thing as a lever and fulcrum. They lessen the force required to move an object by reducing the distance the object needs to move by doubling the rope used. Block, tackle, and pulleys are used by sailors to hoist ropes and sails on ships and by farmers to haul hay bales up into haylofts. I won't go further into how to set them up. I'm no expert. You can get books and videos on how they operate from libraries and websites.

CHAPTER TWELVE

Extras

We live in the modern world, and even though homesteading seems like stepping into the past, into the steps of pioneering forefathers, there is nothing wrong with utilizing modern conveniences. One convenience that comes with lots of benefits is Wi-Fi.

Wi-Fi Accessibility

If Wi-FI is important to you, look into the availability of Wi-Fi in the prospective homesite area. Does your area have service? Will you be using a cell phone or hooking up to the Internet? How much does it cost? How fast is the service?

Some alternative sources for wifi are:

- libraries
- fast food restaurants
- coffee shops
- businesses-Lowe's, Wally World

Wi-Fi is important if you start a YouTube channel for uploading your videos. If you have decided to use only the Wi-Fi in town and keep your land Wi-Fi free, you can still upload at public wifi hotspots. Service on your phone allows you to do banking, online shopping, take pictures and videos, send and receive calls, and perform other modern uses.

CHAPTER THIRTEEN

And Finally... 25 Homesteading Hints

1. Plant trees, especially fruit trees, NOW.
2. Save any baling wire or twine you get with hay/straw bales. You can use it later.
3. Weed and mulch early.
4. Harvest your garden produce when it's ripe. It will not get better with age. It will rot.
5. Older goats have longer horns.
6. Older chickens will have pale legs and blunted beaks.
7. Keep mosquitoes down by reducing standing water and encouraging the growth of bats, toads, frogs, lizards, and birds.
8. Horse worming medicine is the same medicine that is used for dogs, goats, and sheep. It's much cheaper than buying the specialized wormer for dogs, etc. Horse wormers are more concentrated, so they use much less. A tube is in 100-pound weight increments, so for a goat weighing 50 pounds, you'd use 1/2 of one increment. If unsure of the dosage, ask a smarter friend to help you.
9. Provide plenty of space for animals. Overcrowding and unclean conditions bring illness.
10. Plant herbs as well as vegetables and fruit. They are often perennials, which means they will come back each year. Many can be used for tea and medicine.
11. Glass containers with lids keep out pantry moths.
12. Dandelions, plantain, curly dock, chicory, clover, lamb's quarter, purslane, and mullein are all good weeds.
13. Sunflowers are good food for you and all your critters. Plant a perennial variety and let some fall to the ground. They will replant and provide renewed food for you each year. Same with milo grain.
14. Peat moss is an antibacterial, odor eater that can be used in the garden, compost, and for tree bark repair.
15. Acorns, which fall from oak trees, can be food for you and your critters, especially goats and turkeys.
16. Poke, another weed, is high in Vitamin A. Some goats eat the leaves, and some birds eat the fruit. The fruit is poisonous for people but good as a natural dye.
17. Leaves are an excellent mulch and goat food.
18. Ash is a valuable homestead commodity.
    - In the garden to improve the soil and discourage bugs

- In compost
- In-ground food storage medium
- Treatment for swallowed poison
- Make lye for soap
- Melts snow and provides traction on ice

19. Cattails are edible. The tops make great fire starters, and they clean the water where they grow.
20. Mullein can be used for diapers, bandages, and padding. Burnt mullein soothes irritated lungs.
21. Solar Dehydrators will also help on marginal land.
    - Can heat water
    - Dry herbs so you can use them as teas or in tinctures
    - Will dry clothes if you make the trays large enough
    - Can be built by you
    - Dried food is easier to store and takes up less room than frozen food
    - Make jerky
    - Dry grass for feed for your animals.
    - Process all garden produce without electricity or power
22. Those ubiquitous fluid containers: IBC Totes
    - 250 gallons
    - A wire cage, removable
    - Sturdy
    - Compatible connections with PVC pipe and plumbing fittings
    - Can be used as a small animal shelter if you make a door for the entrance.
    - Portable
    - Stackable
    - Paint them black for water storage, so no algae grow.
    - Available at Little Debbie outlets for as low as $40.
23. Keep your eyes peeled for wild-growing fruit trees and vines. If you find either, prune the trees and watch how much fruit they bear the following year. Prune wild grape vines and give them a wooden or metal support; they will bear more next year.
24. Rusty nail water is a mordant for doing your dyeing.
25. Vinegar can be made from a quart of water, apple peels, a tsp of sugar, and a couple of tablespoons of Bragg's Apple Cider Vinegar.

When I started writing this booklet, my husband and I had been homesteading for seven years. Now, three years later, we are still living on our off-grid homestead. It is hard work every day, and, to us, it seems as if we make no progress. Yet we do, and each day brings its challenges and rewards. We have learned so much more than we knew when we started. Spending time homesteading has helped us gain confidence with the experience and the ability to help others by sharing what we've learned. We are convinced living agriculturally is how our Heavenly Father intended His creation to live. It's peaceful, less stressful, and fully satisfying. I hope you found this booklet helpful and feel more confident about your decision to go back to the old ways, the pioneering life that our forefathers lived, and thrive there. Shalom.

About the Author

SK Fox was born in Nicosia, Cyprus while both parents worked for the federal government. After her father passed away, her family moved back to the States. She has a BA in English Literature- Cum Laude. SK Fox has been married to her husband, Joe, for 30+ years. They have 3 children. Her professional career includes NRA Certified Firearms Instructor, PADI and NAUI certified diver, Girl Scout leader, Firefighter, EMT, and Pastor's wife. She also has spent the last 10 years as an off-grid, Torah-observant, homesteading, neo-pioneer and YouTube creator.

See more content from SK Fox on YouTube: Sister Kate Shofar Mountain & Viking Preparedness.

www.ingramcontent.com/pod-product-compliance
Lightning Source LLC
Chambersburg PA
CBHW051333120626
46547CB00016B/2523